# BARNS

# BARNS

Introduction by Charles Leik

FRIEDMAN/FAIRFAX
PUBLISHERS

**A FRIEDMAN/FAIRFAX BOOK**

©1999 by Michael Friedman Publishing Group, Inc.

Library of Congress Cataloging-in-Publication data available upon request.

ISBN 1-56799-802-X

Editor: Emily Zelner
Art Director: Jeff Batzli
Designer: Jennifer S. Markson
Photography Editor: Wendy Missan
Production Director: Karen Matsu Greenberg

Color separations by Colourscan Pte Ltd.
Printed in Singapore by KHL Printing Ltd.

1  3  5  7  9  10  8  6  4  2

For bulk purchases and special sales, please contact:
Friedman/Fairfax Publishers
Attention: Sales Department
15 West 26th Street
New York, New York 10010
212/685-6610   FAX 212/685-1307

Visit our website:
http://www.metrobooks.com

The barn was pleasantly warm in the winter when the animals spent most of their
time indoors, and it was pleasantly cool in the summer when the big doors stood
wide open to the breeze....It was the kind of barn that swallows like to build their
nest in. It was the kind of barn that children like to play in.

—E.B. WHITE, *CHARLOTTE'S WEB*

I saw the spiders marching through air,
Swimming from tree to tree that mildewed day
In latter August when the hay
Came creaking to the barn.

—ROBERT TRAILL SPENCE LOWELL, *MR. EDWARDS AND THE SPIDER*

# BARNS

FEW OF US HAVE EVER MILKED A cow in the predawn winter darkness, wrestled bales of newly mown alfalfa on a hot summer afternoon, experienced the solitude of cattle quietly chewing their cud while a winter storm rages outside, found a hidden litter of kittens in the hay loft, mucked out stables in the gloomy mud days of March, or stripped and graded burley leaf for a tobacco auction, and yet the symbol of the barn is alive and powerful in our minds.

Barns are widely recognized features of our agricultural and physical landscape, but in our highly industrialized society most of us have limited first-hand experience with them. They continue to be active in our vocabulary as metaphor—we may say, for example, that something is "as big as a barn"—and they even find their way to our break-fast tables in the form of advertisements on cereal boxes (which frequently feature a barn in an idyllic setting), but they are increasingly irrelevant to our ordinary, present-day existence. Nonetheless, barns symbolize endurance, security, and stability, and they evoke a romanticized way of life. They remind us of our agrarian past, of less complicated times.

Several years ago I attended the raising of a century-old barn. I expected the participants to be a combination of middle-aged rural men and younger "back to the land" types. Instead, I found a cross section of men and women of all ages and professions. It was apparent that most of them were unfamiliar with agriculture and with post-and-beam construction, but they shared a fascination with participating in a barn raising and anticipated the satisfaction

of bringing an ancient structure of hand-hewn beams back to life.

The barn is vernacular architecture. It is a reflection of the people and history of the region. Few of us can determine the age of a barn or its specific purpose at a glance, but we admire the classical proportions, the honest design and sturdy construction, and the use of native materials, and we can imagine how the building represented the aspirations and success of its first owner.

Most of us have an agricultural background somewhere in our history. My grandfather, for instance, built his own barn at the turn of the century. The neighbors had said that my grandfather would "never make the mortgage" when he purchased the farm. The huge red barn rising on the flat terrain was visible proof that he had not only sur-

vived but was prospering. My grandfather built a gambrel roof barn (often erroneously called a "hip roof"), which has two pitches, or slopes, on each side of the peak. That style, along with the round-roof barn, became popular

Despite our distance from the agricultural way of life, barns remain a powerfully evocative symbol. These enduring monuments of rural life have transcended time and have wondrously continued to feed our imagination and spark our curiosity. Their quiet dignity lends them a certain mystique, while at the same time their rustic, homey image conveys a sense of shelter and warmth. Today, alas, the inventory of these magnificent structures is dwindling at an alarming rate. We invite you to join us in celebrating the barn by embarking on a visual journey of this endangered edifice. Let the following photos and your imagination take you back to an era when many of our ancestors tilled the land and the barn was the nexus of their livelihood.

after the turn of the century in certain regions because it allowed more hay storage in the loft than under the older style. Like most practical architecture, barn design reflected the technological context in which it arose.

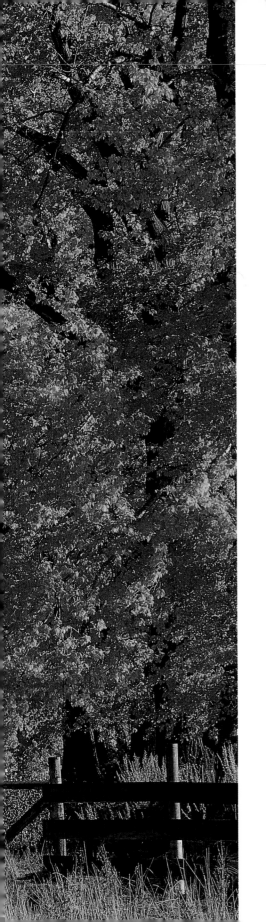

# PHOTO CREDITS